Would You Rather?

Family Challenge!

EDITION

Hilarious Scenarios
& Crazy Competition
for Kids, Teens, and Adults

LINDSEY DALY

Z KIDS • NEW Y

Published in the United States by Z Kids, an imprint of
Zeitgeist™, a division of Random House LLC, New York.

penguinrandomhouse.com

Zeitgeist™ is a trademark of Penguin Random House LLC

ISBN: 9780593435465
Ebook ISBN: 9780593435472

Cover art © Shutterstock.com/Andrii Symonenko
Interior art © Shutterstock.com/mhatzapa
Author photograph by Jocelyn Morales
Book design by Katy Brown

Printed in the United States of America

3 5 7 9 10 8 6 4 2

First Edition

For Mom, Dad, and Andrew, always

&

Mike, Liz, Nicole, Meghan, and Kayla,

my friends who feel like family

Contents

Introduction

My first book, *Would You Rather? Made You Think! Edition,* showed you a fun way to express your creativity in the classroom and with your families. Whether it was in person or over Zoom, you demonstrated just how smart and funny you truly are. You had hilarious, thought-provoking conversations and learned more about your peers through laughter and dialogue. You didn't think I'd stop there, did you? *Would You Rather? Family Challenge! Edition* keeps the critical-thinking party going with over 160 new humorous and interesting scenarios for your whole family to enjoy.

So, let's refresh our memory: What is "critical thinking"? In simple terms, it is considering all of the possibilities around any given topic and forming your own opinion. Much like a toddler who just started talking, thinking critically means learning information and immediately asking, "Why?"

The world NEEDS critical thinkers! Without those skills, Albert Einstein would have never developed the theory of relativity. Martin Luther King, Jr. wouldn't have led the civil rights movement. Malala Yousafzai would have never become an education activist. What do they all have in common? They questioned. They brainstormed. They didn't give up until their thoughts changed the world. They did ALL of their chores without ever being asked! Okay, I can't confirm the last one, but I'm just trying to help your parents out. If these critical thinkers could use their brain power to chase their dreams, so can you!

Start the habit of critical thinking now. Whether you're book smart, street smart, art smart, or all three, this is your time to shine!

Rules of the Game

FOR THREE PLAYERS OR MORE:

* Get a group of friends or family members together for a game of wits and creativity. The more the merrier!
* The game is played in 8 rounds with 20 questions in each round.
* Players rotate the responsibility of being the "judge" and read the question aloud to the group.
* Players will have one minute to develop their answer with an explanation. Players will take turns sharing their answers.
* The judge of that round will then select the answer they enjoyed the most based on humor, creativity, or logic. The player whose answer was chosen will be awarded the point for that question.

* When players complete the round, points will be tallied and someone will be named the winner for that round.
* When players complete the entire book, points will be tallied again to determine the champion.
* In the event of a tie, see the "tie breaker" question at the end of each round. This should only be used for the two players who are tied. All of the remaining players will vote on the best answer.

FOR TWO PLAYERS:

* If you promise to be fair, you can follow the same rules used for three or more players, with both players taking turns at being the judge.
* If not, then each player is given 5 points to award per question. Both players must assign the 5 points in 3 categories: logic, humor, and creativity. Add up the points at the end of each round to see who wins that round!

ROUND 1

Super Science

Would you rather

spend a year working in a top-secret science lab

or

on the International Space Station?

WINNER: POINTS:

Would you rather

dissect a scorpion

or

a tarantula?

WINNER: POINTS:

Would you rather

invent a spaceship that could travel to Mars

or

a device that rids the air of pollutants?

WINNER: POINTS:

Would you rather

discover the bones of the world's first human

or

the world's oldest living reptile?

WINNER: POINTS:

Would you rather
create an app that locates missing pets
or
predicts the winning lottery numbers?

WINNER: POINTS:

Would you rather
chase a tornado in a car
or
stand on top of a skyscraper during a lightning storm?

WINNER: POINTS:

Would you rather

spend a year studying volcanic ash

or

the Earth's crust?

WINNER: POINTS:

Would you rather

watch a solar eclipse

or

a meteor shower?

WINNER: POINTS:

Would you rather
have a planetarium
in your home
or
a greenhouse full of
exotic plants?

WINNER: POINTS:

Would you rather
develop a vaccine that prevents
people from itching
or
stops people from feeling cold?

WINNER: POINTS:

Would you rather

be fluent in writing HTML code

or

know how to hack your favorite video game so you always win?

WINNER: POINTS:

Would you rather

spend an hour floating around in space

or

swimming in the depths of the ocean?

WINNER: POINTS:

Would you rather
have an arm made of gold
or
aluminum?

WINNER: POINTS:

Would you rather
examine parasites
under a microscope
or
a new strain of bacteria?

WINNER: POINTS:

Would you rather
discover a new element
or
a new mineral?

WINNER: POINTS:

Would you rather
save Earth from plummeting
into a black hole
or
being hit by a large meteor?

WINNER: POINTS:

Would you rather
discover a faster method of air travel
or
land travel?

WINNER: POINTS:

Would you rather
research plants in the
Amazon rain forest
or
rock formations in the
Grand Canyon?

WINNER: POINTS:

Would you rather

own a star

or

a piece of coral from the Great Barrier Reef?

WINNER: POINTS:

Would you rather

accidentally set a science lab on fire during an experiment

or

get a dangerous chemical in your eye?

WINNER: POINTS:

TIEBREAKER

Would you rather

watch how fireworks are made

or

jet packs?

WINNER: POINTS:

WINNER: ______________________________

TOTAL POINTS: ______________________________

ROUND
2

Nifty Nature

Would you rather
have a pet owl
or
giraffe?

WINNER: POINTS:

Would you rather
watch baby birds hatching
or
baby sea turtles?

WINNER: POINTS:

Would you rather

get thrown up on by your dog

or

licked by your cat after it just cleaned itself?

WINNER: POINTS:

Would you rather

play on a rope swing over a creek

or

go white water rafting down a river?

WINNER: POINTS:

Would you rather

raise a baby bunny that has been separated from its mother

or

nurse a bird with a hurt wing back to health?

WINNER: POINTS:

Would you rather

walk through a spider web

or

a pricker bush?

WINNER: POINTS:

Would you rather
play hide-and-seek
in a dense forest
or
a cornfield with cornstalks
taller than your head?

WINNER: POINTS:

Would you rather
get knocked over by a giant wave
or
caught outside in a hail storm?

WINNER: POINTS:

Would you rather
eat a live toad
or
a dead rattlesnake?

WINNER: POINTS:

Would you rather
fall face-first into a
hornet's nest
or
a pile of horse manure?

WINNER: POINTS:

Would you rather

hike ten miles in the pouring rain

or

bike five miles in the sweltering heat?

WINNER: POINTS:

Would you rather

use wet leaves as toilet paper

or

tree bark?

WINNER: POINTS:

Would you rather

spend a night in
an igloo wearing only
your bathing suit

or

the middle of a forest
during a thunderstorm?

WINNER: POINTS:

Would you rather

give a piggyback ride to a penguin

or

a sloth?

WINNER: POINTS:

Would you rather
build a shelter using cacti
or
bamboo sticks?

WINNER: POINTS:

Would you rather
prevent wildfires
or
eliminate all of the trash in the world's oceans?

WINNER: POINTS:

Would you rather

be caught in a volcanic eruption

or

an avalanche?

WINNER: POINTS:

Would you rather

spend the day playing in an overgrown meadow

or

a tree house?

WINNER: POINTS:

Would you rather
wake up covered in centipedes
or
worms?

WINNER: POINTS:

Would you rather
be a stink bug for a week
or
a cockroach?

WINNER: POINTS:

TIEBREAKER

Would you rather

hug a jellyfish

or

a stingray?

WINNER: POINTS:

WINNER: ______________________________

TOTAL POINTS: ______________________________

ROUND
3

Serious Sports

Would you rather

get hit with a foul ball
at a baseball game

or

tackled by a football player?

WINNER: POINTS:

Would you rather

play your personal best but your team loses

or

play the worst game of your life but your team wins?

WINNER: POINTS:

Would you rather
compete in a swimming race
wearing a gorilla costume
or
a hazmat suit?

WINNER: POINTS:

Would you rather
ride in the passenger seat
of a car in a NASCAR race
or
on the back of a horse in
the Kentucky Derby?

WINNER: POINTS:

Would you rather
be the caddy for a professional golfer
or
the ball person during
a tennis match?

WINNER: POINTS:

Would you rather
win front-row tickets
to the Super Bowl
or
play a one-on-one game against
your favorite athlete?

WINNER: POINTS:

Would you rather

be thrown into a skateboarding competition without any experience

or

a surfing competition?

WINNER: POINTS:

Would you rather

be a professional polo player

or

bowler?

WINNER: POINTS:

Would you rather

race someone on a Jet Ski

or

a dirt bike?

WINNER: POINTS:

Would you rather

wrestle someone 50 pounds heavier than you

or

face professional players in a rugby match?

WINNER: POINTS:

Would you rather

make the game-winning
basket at the buzzer

or

hit a home run in the final inning?

WINNER: POINTS:

Would you rather

be the best player on a horrible team

or

the worst player on the best team?

WINNER: POINTS:

Would you rather

compete in an archery tournament

or

a bobsled race?

WINNER: POINTS:

Would you rather

be chosen team captain over your best friend

or

have your best friend be picked instead of you?

WINNER: POINTS:

Would you rather

your team lose a championship game by 1 point

or

50 points?

WINNER: POINTS:

Would you rather

have an ice-skating rink in your backyard

or

an Olympic-size swimming pool?

WINNER: POINTS:

Would you rather

be a referee for professional sporting events

or

a commentator?

WINNER: POINTS:

Would you rather

play beach volleyball when it's snowing

or

100 degrees?

WINNER: POINTS:

Would you rather

be a flyer at the top of a cheerleading pyramid

or

on a synchronized ice-skating team?

WINNER: POINTS:

Would you rather

be a competitive dancer

or

an acrobat in a traveling circus?

WINNER: POINTS:

TIEBREAKER

Would you rather

run a marathon in snow boots

or

fancy dress shoes?

WINNER: POINTS:

WINNER: ______________________________

TOTAL POINTS: ______________________________

ROUND 4

Hilarious History

Would you rather
discover the diary
of a Roman emperor
or
a Founding Father
of the United States?

WINNER: POINTS:

Would you rather
have been in the audience
for Martin Luther King, Jr.'s
"I Have a Dream" speech
or
Abraham Lincoln's
"Gettysburg Address"?

WINNER: POINTS:

Would you rather

have a one-on-one lesson
with Socrates

or

Albert Einstein?

WINNER: POINTS:

Would you rather

be able to go back in time and
change one historical event

or

see 200 years into the future but
not be able to change anything?

WINNER: POINTS:

Would you rather

fall down the stairs
of the Lincoln Memorial

or

get stuck in the elevator inside
the Washington Monument?

WINNER: POINTS:

Would you rather

be a member of the CIA

or

the Secret Service?

WINNER: POINTS:

Would you rather

accidentally break an ancient vase in a museum

or

damage the original US Constitution by touching it?

WINNER: POINTS:

Would you rather

have a sleepover with friends in a famous museum

or

the White House?

WINNER: POINTS:

Would you rather

find out that you're a descendant of a former US president

or

a medieval king?

WINNER: POINTS:

Would you rather

interview a military general

or

an FBI agent?

WINNER: POINTS:

Would you rather

be a history teacher

or

a museum tour guide?

WINNER: POINTS:

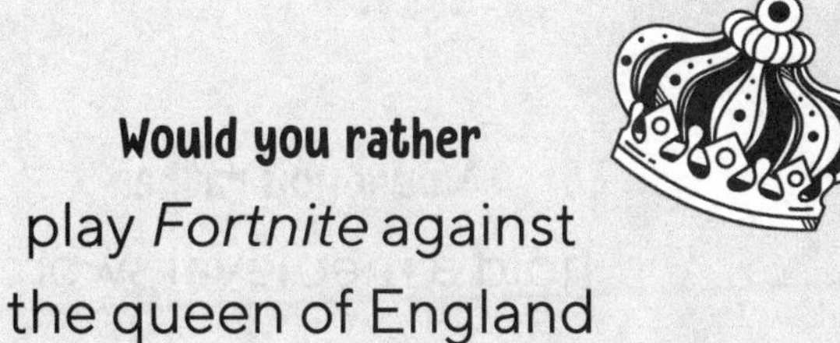

Would you rather

play *Fortnite* against the queen of England

or

Among Us against the prime minister of Canada?

WINNER: POINTS:

Would you rather
be a presidential speechwriter
or
a member of Congress?

WINNER: POINTS:

Would you rather
hold a medieval sword
or
try on an ancient Egyptian headdress?

WINNER: POINTS:

Would you rather

be a court jester

during the Middle Ages

or

the king's food taster?

WINNER: POINTS:

Would you rather

be awarded the Nobel Peace Prize

or

the Presidential Medal of Freedom?

WINNER: POINTS:

Would you rather

be a Supreme Court justice

or

a United Nations ambassador?

WINNER: POINTS:

Would you rather

be a cowboy or cowgirl

in the Wild West

or

a jazz musician during the

Roaring Twenties?

WINNER: POINTS:

Would you rather
spend the night alone
on a former battlefield
or
in Arlington National Cemetery?

WINNER: POINTS:

Would you rather
accidentally sneeze in the
face of a famous activist
or
the president of the United States?

WINNER: POINTS:

Would you rather

help repair the Great Wall of China

or

the Great Pyramids?

WINNER: POINTS:

WINNER:

TOTAL POINTS:

ROUND
5

Essential Entertainment

Would you rather
be the star of a Broadway show that nobody liked
or
the author of a book that got bad reviews?

WINNER: POINTS:

Would you rather
spend a day shadowing an animator
or
a CGI (computer-generated images) artist?

WINNER: POINTS:

Would you rather

be pranked by a famous YouTube star

or

surprised by your favorite celebrity?

WINNER: POINTS:

Would you rather

be an extra in a film

or

a stunt double?

WINNER: POINTS:

Would you rather

take a six-hour mime class

or

perform magic tricks at a five-year-old's birthday party?

WINNER: POINTS:

Would you rather

be a talented actor that few have heard of

or

an average actor known by all?

WINNER: POINTS:

Would you rather
be a backup dancer
for a famous singer
or
a news reporter?

WINNER: POINTS:

Would you rather
star in a reality television show
that documents your life
or
be the host of a popular podcast?

WINNER: POINTS:

Would you rather
try stand-up comedy
and completely bomb
or
have your voice crack while
singing for an audience?

WINNER: POINTS:

Would you rather
represent celebrities
as their publicist
or
lawyer?

WINNER: POINTS:

Would you rather

be prevented from watching your favorite movie again

or

listening to your favorite song?

WINNER: POINTS:

Would you rather

get a private tour of the set of your favorite television show

or

have a five-minute conversation with your favorite actor?

WINNER: POINTS:

Would you rather

learn how to perform stage combat

or

make elaborate costumes?

WINNER: POINTS:

Would you rather

do a dramatic poetry reading in front of your neighbors

or

perform a tap-dancing routine?

WINNER: POINTS:

Would you rather

tour the world as a member
of a ten-person band

or

a solo artist?

WINNER: POINTS:

Would you rather

host an awards show

or

interview celebrities on
the red carpet?

WINNER: POINTS:

Would you rather

get paid to play video games

or

review books?

WINNER: POINTS:

Would you rather

entertain children at an amusement park dressed as an animal

or

a clown?

WINNER: POINTS:

Would you rather

go crowd-surfing during a concert

or

dance onstage?

WINNER: POINTS:

Would you rather

be a contestant on a game show that tests your mental abilities

or

your physical strength?

WINNER: POINTS:

TIEBREAKER

Would you rather

attend a concert where the only instrument played is an accordion

or

a tambourine?

WINNER: POINTS:

WINNER: ______________________

TOTAL POINTS: ______________________

ROUND
6

Genius Geography

Would you rather
spend a summer traveling
to every city in Europe
or
Asia?

WINNER: POINTS:

Would you rather
draw maps for a living
or
be an archaeologist?

WINNER: POINTS:

Would you rather

hike every mountain
in North America

or

kayak across all of the Great Lakes?

WINNER: POINTS:

Would you rather

live in an isolated cabin for a year

or

the world's most crowded city?

WINNER: POINTS:

Would you rather
own an island
or
a private lake?

WINNER: POINTS:

Would you rather
spend a month living in
a tent in Antarctica
or
the Sahara Desert?

WINNER: POINTS:

Would you rather

sail across the Atlantic Ocean

or

walk the entire length of California?

WINNER: POINTS:

Would you rather

go zip-lining over the Arenal Volcano in Costa Rica

or

skydiving in Sydney, Australia?

WINNER: POINTS:

Would you rather

go snow tubing down Mt. Everest

or

roller-skating on the Great Wall of China?

WINNER: POINTS:

Would you rather

go snorkeling in the Great Barrier Reef

or

rock climbing at Yosemite National Park?

WINNER: POINTS:

Would you rather
closely examine a glacier
or
lava on a dormant volcano?

WINNER: POINTS:

Would you rather
take aerial pictures of rivers
or
photograph underground caverns?

WINNER: POINTS:

Would you rather

spend one day in every country in the world

or

visit 20 countries but spend as much time as you wanted exploring them?

WINNER: POINTS:

Would you rather

climb to the top of the Eiffel Tower

or

Machu Picchu?

WINNER: POINTS:

Would you rather

go surfing down a huge sand dune

or

relax in a hot spring?

WINNER: POINTS:

Would you rather

go on an African safari

or

visit an elephant sanctuary in Thailand?

WINNER: POINTS:

Would you rather

tour Brazil by helicopter

or

London from the top of a double-decker bus?

WINNER: POINTS:

Would you rather

study the effects of hurricanes

or

droughts?

WINNER: POINTS:

Would you rather

discover a way to prevent glaciers from melting

or

earthquakes from happening?

WINNER: POINTS:

Would you rather

watch the sunrise from a beach in Hawaii

or

the sunset from the top of Mount Fuji?

WINNER: POINTS:

TIEBREAKER

Would you rather

find your way through a vast forest using only a compass

or

a poorly drawn map?

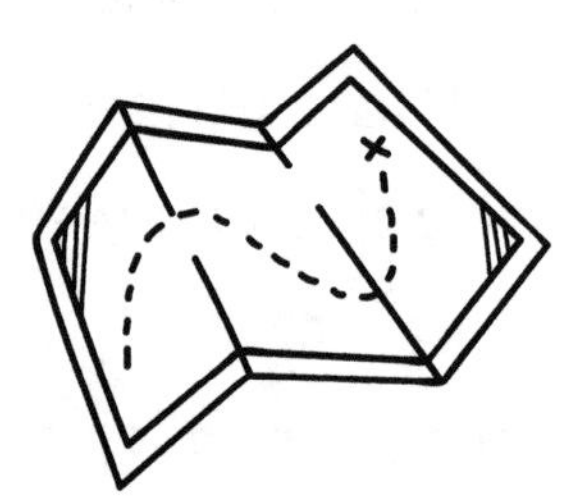

WINNER: POINTS:

WINNER: ____

TOTAL POINTS: ______________________________________

ROUND 7

Astonishing Arts

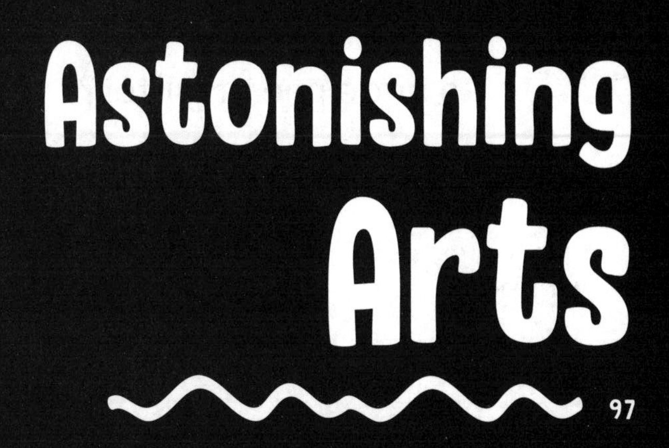

Would you rather

be given the *Mona Lisa*
by Leonardo da Vinci

or

Starry Night by Vincent van Gogh?

WINNER: POINTS:

Would you rather

jump into a pool filled with paint

or

a huge pile of glitter?

WINNER: POINTS:

Would you rather

watch Shakespeare perform
one of his plays

or

listen to Mozart play piano?

WINNER: POINTS:

Would you rather

work in an art gallery

or

be an art therapist for kids?

WINNER: POINTS:

Would you rather
compose a symphony
or
choreograph a ballet?

WINNER: POINTS:

Would you rather
take a class on graffiti painting
or
glassblowing?

WINNER: POINTS:

Would you rather
learn how to play
the violin
or
the tuba?

WINNER: POINTS:

Would you rather
dance in a flash mob
or
sing with an a cappella group?

WINNER: POINTS:

Would you rather

sit perfectly still for two hours to pose for a sculpture class

or

run and jump constantly for an exercise photo shoot?

WINNER: POINTS:

Would you rather

be a makeup artist for a horror film

or

a set designer?

WINNER: POINTS:

Would you rather
be a jeweler
or
make wax figures?

WINNER: POINTS:

Would you rather
create art out of recycled materials
or
food products?

WINNER: POINTS:

Would you rather

be a famous fashion designer of baby clothing

or

pet accessories?

WINNER: POINTS:

Would you rather

take a private dance lesson with Misty Copeland

or

a writing workshop with Suzanne Collins?

WINNER: POINTS:

Would you rather
have one of your designs featured during New York Fashion Week
or
one of your paintings shown in a museum?

WINNER: POINTS:

Would you rather
teach a group of senior citizens how to do a TikTok dance
or
beatbox?

WINNER: POINTS:

Would you rather

learn the art of printmaking

or

calligraphy?

WINNER: POINTS:

Would you rather

be an interior designer

or

a film director?

WINNER: POINTS:

Would you rather

have your face covered
in papier-mâché

or

acrylic paint?

WINNER: POINTS:

Would you rather

be a tattoo artist

or

an illustrator for children's books?

WINNER: POINTS:

TIEBREAKER

Would you rather

have the ability to see a color that no one else can see

or

hear a musical note that no one else can hear?

WINNER: POINTS:

WINNER: ________________________

TOTAL POINTS: ________________________

ROUND
8

Fabulous Fantasy

Would you rather
anger Poseidon
or
Zeus?

WINNER: POINTS:

Would you rather
discover a forest where candy grows on trees
or
live in a house where the furniture is edible?

WINNER: POINTS:

Would you rather

have a conversation with an alien

or

a robot that can no longer be controlled by the humans who created it?

WINNER: POINTS:

Would you rather

be able to change your voice to mimic whoever you want

or

run twice as fast as the world's fastest person?

WINNER: POINTS:

Would you rather

find the lost city of Atlantis

or

the Fountain of Youth?

WINNER: POINTS:

Would you rather

be able to charge your electronic devices with the snap of your fingers

or

generate Wi-Fi by clapping your hands?

WINNER: POINTS:

Would you rather

have the opportunity to feed a live dinosaur

or

ride on a woolly mammoth?

WINNER: POINTS:

Would you rather

have chocolate syrup come out of your eyes when you cry

or

soda?

WINNER: POINTS:

Would you rather
be abducted by a friendly group of aliens for a week
or
travel back in time to spend a week with cave people?

WINNER: POINTS:

Would you rather
summon a ghost with a Ouija board
or
meet a leprechaun?

OUIJA

WINNER: POINTS:

Would you rather

have a genie grant you three wishes immediately

or

five wishes over the span of five years?

WINNER: POINTS:

Would you rather

have a pet that can understand and speak your language for one full day

or

be able to speak to wild animals whenever you want?

WINNER: POINTS:

Would you rather

get stuck in an alternate universe where people sing to express their feelings

or

people are incapable of lying?

WINNER: POINTS:

Would you rather

drink a magic potion that gives you the ability to walk through walls

or

pause time?

WINNER: POINTS:

Would you rather

have confetti fly out of your nose every time you sneeze

or

sugar?

WINNER: POINTS:

Would you rather

have the ability to breathe underwater

or

be invisible for one day each month?

WINNER: POINTS:

Would you rather

never need to sleep

or

use the bathroom?

WINNER: POINTS:

Would you rather

have a fire-breathing dragon as a pet

or

a werewolf?

WINNER: POINTS:

Would you rather

be taken captive by
a group of
garden gnomes

or

fairies?

WINNER: POINTS:

Would you rather

wake up on a train with no conductor

or

a bus driven by a zombie?

WINNER: POINTS:

TIEBREAKER

Would you rather

walk through a secret door that takes you back in time to a medieval European town

or

a colonial American city?

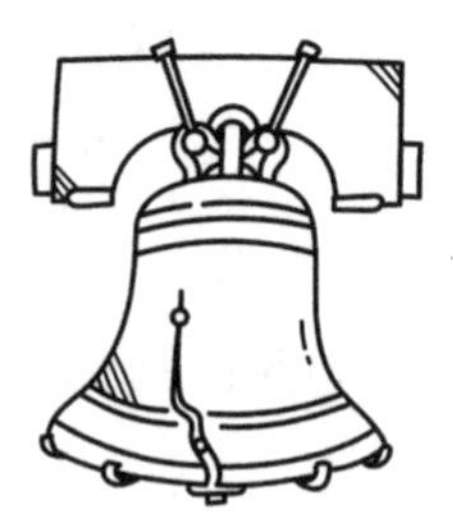

WINNER: POINTS:

WINNER: ___

TOTAL POINTS: ___

WINNER
Would You Rather?
FAMILY CHALLENGE!
EDITION

This certificate
is awarded to

---

for being ferociously funny,
captivatingly creative, super smart,
and the winner of wit.

A person who can explain AND
entertain!

CONGRATULATIONS!

About the Author

Lindsey Daly grew up in Andover, New Jersey. She graduated from Ramapo College of New Jersey with a BA in history and a certification in secondary education. Lindsey is a middle school social studies teacher and the author of *Would You Rather? Made You Think! Edition*. She lives with her dog, Teddy, in New Jersey.

Instagram: **@lindseydalybooks**
Twitter: **@LindseyDaly10**